BUILDING BLOCKS OF COMPUTER SCIENCE

CYBERSECURITY

Written by Echo Elise González

Illustrated by Graham Ross

WORLD BOOK

www.worldbook.com

Co-published by agreement between Shi Tu Hui and World Book, Inc.

Shi Tu Hui
Room 1807, Block 1,
#3 West Dawang Road
Chaoyang District, Beijing 100025
P.R. China

World Book, Inc.
180 North LaSalle Street
Suite 900
Chicago, Illinois 60601
USA

Library of Congress Control Number: 2025938328

Building Blocks of Computer Science
ISBN: 978-0-7166-6706-3 (set, hard cover)

Cybersecurity
ISBN: 978-0-7166-6716-2 (hard cover)

Also available as:
ISBN: 978-0-7166-6726-1 (soft cover)
ISBN: 978-0-7166-6736-0 (e-book)

WORLD BOOK STAFF

Editorial

Vice President
Tom Evans

Senior Manager, New Content
Jeff De La Rosa

Manager, New Product Development
Nicholas Kilzer

Associate Manager, New Content
William D. Adams

Content Creator
Elizabeth Huyck

Proofreader
Nathalie Strassheim

Graphics and Design

Senior Visual Communications Designer
Melanie Bender

Acknowledgments
Writer: Echo Elise González
Illustrator: Graham Ross/The Bright Agency
Series reviewed by George K. Thiruvathukal (Loyola University Chicago); Peter Jang (Actualize Coding Bootcamp)

TABLE OF CONTENTS

There is a glossary on page 39. Terms defined in the glossary are in type **that looks like this** on their first appearance.

WHAT IS CYBERSECURITY?
Hello there, young citizen of the digital world!
I'm Guard! I'm a **firewall**, a type of computer security system.
FIREWALL
I'm here to tell you about one of the most important topics of our time: **cybersecurity**.
CYBERSECURITY

Look around you. Do you see any computers near you right now?
Any laptops? Phones? Smart screens? Are you reading this book on a tablet?
Computer systems are all around us. They help us communicate, learn, create, and even stay healthy.
People all over the world rely on computers, computer systems, and computer networks in many aspects of their lives.
That's why it's so important to keep computer systems as safe and secure as possible.

That's where I come in! As a security system, my job is to protect the digital world from harm.
Hey, are you OK, Flow?
Yeah, I'm over here!
Wait! You're my arch-nemesis, GLITCH!

Glitch is a **Trojan horse,** a type of computer virus.
A Trojan horse sneaks into a computer system by appearing to be a useful and nonharmful program.
But, really, it contains malicious code that threatens the security of the system! Stay away, Glitch!
A Trojan horse attack is just one of many different kinds of cyberattacks.
CYBERATTACK
Cyberattacks are a way for criminals to get control of people's phones and computers.
Fortunately, cybersecurity prevents many attacks. And when an attack does happen, we can use cybersecurity methods to minimize the damage.

CYBERCRIMINALS ON THE LOOSE!
Cybercriminals are people who use computers to commit crimes like cyberattacks.
Why do people commit cyberattacks?
Sometimes they do it to steal money or data from another person or organization.
BREAKING NEWS
ALL SYSTEMS ARE MYSTERIOUSLY DOWN
Sometimes they do it to disrupt an organization or prevent a system from working properly.

Cyberattacks can cause great harm to people and organizations.
They can result in identity theft...
STATEMENT
Hey, I didn't buy that!
...and the loss of money or data.
Our files!
Ugh, so slow.
Sometimes a cyberattack creates a small inconvenience for the victim.
OH NO!!!
Other times, it can be disastrous or even life-threatening.

Every year, thousands of cyberattacks are committed.
They affect hundreds of millions of people!
To successfully combat cyberattacks, it's important that we understand exactly what we are up against.
There are many different types of threats to cybersecurity.
One of the most common threats comes in the form of malware.
Malware stands for MALicious softWARE.
Malware is **software** that is designed to disrupt or damage a computer or computer system.
MALWARE
MALICIOUS SOFTWARE
Our adversary, Glitch the Trojan horse, is an example of one kind of malware.

Other malwares include spyware, ransomware, worms, and botnets.
Spyware is software that secretly gathers information from a computer and sends it back to a cybercriminal.
Ransomware locks up computer files or systems until ransom money has been paid.
Pay UP
A worm is a kind of malware that can replicate itself and spread to other devices on its own.
A botnet is a network of infected devices that are all connected via the internet.
Cybercriminals use botnets to make organized, massive attacks.

EMAIL
CLICK FOR CUTE PONY PICS
CLICK
Some cybercriminals trick people into sending them money or personal data.
CLICK
They may send an email that has a link in it.
That link could take you to a fake website that collects your personal information.
Sending malicious emails like this is called **phishing.**
Pretending to be a different person or organization is called **spoofing.**
Phishing and spoofing are two of the most common ways that cybercriminals attack.
PHISHING
SPOOFING
GUARD'S CYBER SAFETY TIPS
Never click on a link from an email or text message, unless you're 100% sure who sent it to you and where it goes!

A **computer hacker** is a cybercriminal who breaks into computer systems.
ACCOUNT
PASSWORD
GERBIL4
Some hackers try to crack passwords to access restricted sites, devices, and networks.
BRUTE FORCE ATTACK
Guessing passwords until you hit on the right one is called a brute force attack.
Usually, hackers use software to cycle through millions of passwords until they find the correct one.
PASSWORD
ACCESS GRANTED
HAMSTER111
GUARD'S CYBER SAFETY TIPS
Long, random passwords are the safest! Use a long phrase or a mix of lowercase letters, uppercase letters, numbers, and symbols. Don't use words that are easy to guess.

Instead of trying to crack passwords, some hackers conduct man-in-the-middle (MITM) attacks.
CAFE
MITM
In this kind of attack, the criminal gets in the middle of a digital communication between two people.
password
account number
$1,000
1234 Rabbit Street
This allows them to intercept private information, such as passwords and account numbers.
If a person does online banking using public Wi-Fi, they risk having their information stolen by any cybercriminals who might be spying on devices using that public Wi-Fi network.
GUARD'S CYBER SAFETY TIPS
Only access personal accounts and sensitive data using secure, private Wi-Fi and cellular networks.

CYBERATTACK SPECIALS
DoS
Another kind of attack tries to keep users from accessing specific sites or networks. These are called DoS (denial-of-service) attacks.
Sometimes, a DoS attack involves flooding a server (a computer that is a hub for a network of other computers) with more internet traffic than it can handle.
SITE UNAVAILABLE
SITE UNAVAILABLE
The server shuts down and then no one can access its services.
CYBERATTACK SPECIALS
DoS
DDoS
This is called a DDoS (distributed denial-of-service) attack because the flood of internet traffic comes from multiple devices at once.

PREVENTING CYBERATTACKS
As you can see, there are many digital threats that we have to look out for.
There are dangerous foes galore!
But fear not! Cybersecurity is here to save the day!
CYBER SECURITY
Cybersecurity is a powerful set of tools that help protect us from cyberattacks.
Let's see how they work!

LIBRARY
The first step in staying safe from cyberattacks is PREVENTION!
Cybersecurity experts have developed many ways to prevent cyberattacks before they start. Take me, for example. I'm a firewall.
I safeguard computers and networks that use the internet, especially ones that use public Wi-Fi.
10101
01
When any data tries to enter your network or computer, I scan it.
10101
010
If I see anything suspicious, I block it from accessing your computer!

I want to send a top-secret message to my buddy, Algorithm.
But how can I make sure no hackers will be able to read it?
To: Al@chipset.mycomputer
Hi Al.
Here is my payment information, so I can chip in for Chips birthday gift.
12345-67890
SEND
ENCRYPTION
One way is to use **encryption.** Encryption turns the data that makes up my message into a special code so that hackers can't read it.
SEND
Hi Al
Hi Al
A properly encrypted code can only be deciphered by the program that is sending the data and the program that is meant to receive it.

MULTIFACTOR AUTHENTICATION
Many websites and applications use **multifactor authentication,** or MFA, to keep the data safe and prevent cyberattacks.
HAPPY BIRTHDAY
With MFA, I need to take extra steps to confirm my identity.
In order to access Chip's birthday party, I have to scan my hand AND enter this code that was texted to me.
CODE 6RJ75 #279
Hmm... It looks like someone is trying to use my identity to gain access.
HERE IS YOUR NEW CODE 22761
It's a good thing Chip's house has MFA!
Aha! Glitch tried to sneak in, but the access code was sent to MY phone.
MFA creates an extra step, but I don't mind because it helps us protect against cybercriminals!

DETECTING CYBERATTACKS
Even a strong firewall like myself can't always catch EVERY sneaky piece of malware.
Mistakes can happen! Sometimes passwords get shared.
And cybercriminals are always thinking of new ways to attack people's computers.
When a cyberattack does happen, we must be able to recognize that it is happening so we can address the problem.

One clever way to both prevent and detect a cyberattack is by setting up a honeypot.
Not that kind of honey pot!
HON
I mean a decoy computer that mimics other computers in a network.
This honeypot blends in with the other computers in our network...
CYBERATTACK TRACKER
PERSONAL ACCOUNT INFORMATION
$3,500.00
$1,000.00
$555.00
$2,679.00
But it contains fake data to lure in cybercriminals.
It is equipped with trackers and other software that help us detect cyberattacks before they affect our real data.
CYBERATTACK TRACKER
Everyone, turn off your Wi-Fi connection so Glitch cannot spread to the real computers!
CYBERATTACK TRACKER
Honeypots detect cyberattacks and help us learn about them.

This is my cousin, AV. She's an **antivirus** program.
Antivirus software scans your computer's memory for signs of malware.
AV can identify many types of malware and block them from harming your computer.
I work on the front lines of the internet.
But, if something slips past me, AV can still provide a good line of defense for your computer!

Many antivirus programs work with **artificial intelligence**, or AI, to identify threats.
AI programs can learn from data and adapt without receiving specific instructions from a programmer.
Let's adjust our programming so that spyware can't get in again.
Antivirus programs that use AI learn to recognize the data patterns that indicate the presence of malware.
WANTED
REWARD
WANTED
REWARD
This allows them to recognize new threats, and to learn how to detect threats better. Fantastic!

SYSTEM RECOVERY
Oh No! Despite our best efforts, Glitch has found a way into our system!
It's a good thing we have a system recovery plan in place.
SYSTEM RECOVERY PLAN
First things first... We've got to contain the threat so it doesn't spread too far!
We've isolated the infected program from the rest of the computer...
Now, Glitch won't be able to attack the rest of our system.

Now that Glitch is contained, AV can remove it.
This is the second step in our system recovery plan. No more horsing around here, Glitch!
Now we can figure out what vulnerabilities allowed Glitch to infiltrate, and we can patch them.
It looks like Glitch got in through an email.
We should update our email software to make sure we have the latest version.
I'll remind everyone to be extra careful not to click on any suspicious email links.
Fortunately, we backed up all the data that Glitch was able to steal.
DATA BACKUP

CYBER HEROES!
Cybersecurity is a team effort.
I'd like to introduce you to... the League of Cyber Heroes!
These are people who work hard every day to protect folks from cyberattacks.
A=8y7
B=k4-
C=+05
I'm an encryption expert! I develop cryptographic code to keep data safe from hackers.
I also test cryptographic security systems to make sure they don't have any weaknesses or vulnerabilities.
Encryption experts are also called **cryptographers**.
K3*G!DmM6%

I'm an **ethical hacker.** I know a lot of the same techniques that criminal hackers use to commit cyberattacks.
BUT I use them to help people, rather than harm them!
GLITCH
Ethical hackers are really important.
They help us see inside the minds of hackers, so we can anticipate their attacks.
You have to be creative and crafty to be a good ethical hacker!

def <square numbers(numbers)
""""Return a list of the squares of the numbers
squares=[ ]
for number in numbers
squares.append (number **/)
return squarcs
def main():
numbers=[1,2,3,4]
result=square_numbers(numbers)
Print (result)
I'm a security software developer. I create antivirus programs and other software that protects computers from cyberattacks.
Cybercriminals are always thinking of new ways to get past security software...
...So I must keep making stronger and better programs. I love coding, so this is a perfect job for me!
TAP TAP
Security software developers create some of the most important computer programs in the world!

Oh, hi there! What's your job?
I'm a threat hunter!
BONK!
I look for cybersecurity threats, especially those that would be hard or impossible for software to detect.
DOWNLOADING DATA
For example, this office worker is using his special access to steal data from the company he works for.
He's not using any malware, so an antivirus program won't be able to catch him...
But I can!
Threat hunters are good at noticing patterns and analyzing data to understand where threats are coming from.

Do you think you have what it takes to join the League of Cyber Heroes?
We think you definitely do!!
No matter what your strengths and interests are, there's a place for YOUR special powers in the world of cybersecurity.
Lots of people who work in cybersecurity have a degree in computer science, computer engineering, mathematics, or in cybersecurity itself.
But cybersecurity professionals come from all different kinds of backgrounds! And they all play important roles in keeping the computers of the world safe.

Everyone can take simple steps to prevent cyberattacks.
So, remember!
STRONG PASSWORDS
Make good, strong passwords!
DIFFERENT PASSWORDS
Don't use the same password for different accounts!
PRIVATE PASSWORDS
Never share your passwords!
AVOID SUSPICIOUS LINKS
Don't click suspicious links, especially if you don't know the person who sent it.
KEEP YOUR PERSONAL INFORMATION PRIVATE
And don't share personal information online.
Got it? Great!
You're already on your way to being a CYBER HERO!

TIMELINE

1967

The term "computer security" was first used in a document by the United States Air Force, highlighting the increasing importance of protecting computer data.

1972

The Reaper antivirus program was written to counteract the first computer virus, called "Creeper."

1986

The U.S. government passed the Computer Fraud and Abuse Act, which criminalizes unauthorized access to computer systems and networks.

1988

The Morris Worm, the first major computer worm to spread across the internet, infected roughly 6,000 computers worldwide.

1989

The first "ransomware" virus, called the PC Cyborg Trojan, encrypted filenames on the infected computer's hard drive and demanded a payment for the files to be restored.

1991

The PGP (Pretty Good Privacy) encryption program was developed to provide end-to-end security for personal and business communications.

2000

The ILOVEYOU virus spread globally through email, affecting millions of computers and causing significant financial damage.

2004

Introduction of Stateful firewalls, the first modern firewall standards.

2010

A cyberattack using the Stuxnet computer worm crippled the nuclear energy program of Iran, in the first use of cyber warfare.

2017

The WannaCry ransomware attack spread globally, affecting hundreds of thousands of computers across 150 countries.

ACTIVITY: CYBERSECURITY TIPS

Most everyone today has a home computer and smartphone connected to the internet. Keeping your personal data and information secure is more important than ever.

Here are 10 tips to help keep your computer safe from online threats:

1. Use Strong Passwords

Make your passwords tough to guess! Try mixing uppercase and lowercase letters, numbers, and symbols. A password like "Pa$$w0rd123" is way better than "password123." And don't use the same password for everything!

2. Enable Multifactor Authentication (MFA)

It's like a second lock on your door. When you log in, you'll get a code sent to your phone or email. Even if someone figures out your password, they still need that second code to get in.

3. Update Your Software Regularly

Updates might seem annoying, but they're super important! They fix bugs and security holes, so hackers can't sneak in. Always click "Update" when your computer or apps ask for it!

4. Install Antivirus Software

Antivirus programs help protect your computer from bad stuff, such as viruses, malware, and spyware. Think of it like a digital security guard keeping watch.

5. **Avoid Clicking Unknown Links**

If you get an email or message with a link that looks weird or is from someone you don't know, don't click it! It could be a trick to steal your info. Always double-check before you click.

6. **Be Careful on Public Wi-Fi**

Public Wi-Fi is like a free snack—super tempting, but not always safe. Don't do anything sensitive (like banking) when you're on public Wi-Fi unless you're using a VPN (Virtual Private Network) to protect your connection.

7. **Things Happen**

If your computer crashes or gets hacked, having backups means you don't lose your important stuff. Use a cloud service or an external hard drive to keep a copy of your files.

8. **Use a Firewall**

Firewalls act like digital bouncers, blocking unauthorized access to your computer. Most operating systems have one built-in, so make sure it's turned on.

9. **Log Out of Accounts When You're Done**

If you're using someone else's computer, or even your own, always log out of your accounts when you're done. This helps keep your info safe, especially if you're sharing a device.

10. **Think Before You Share**

Keep your personal info private. Don't overshare on social media or in messages. Even something small, like your birthday or pet's name, could be the clue a hacker needs to guess your passwords.

Stay safe out there and keep your computer protected like a pro!

CAN YOU BELIEVE IT?!

The very first computer virus, developed in the 1970's, was called

the "Creeper."

It displayed the message, "I'm the creeper, catch me if you can!" and moved from one computer to another on ARPANET, the early version of the internet.

In 1999, a 13-year-old hacker named Jonathan James (aka "C0d3br34k3r") broke into NASA's computers. He was the

first juvenile

to be sent to jail for hacking into the U.S. government's systems!

The first "Firewall" was

actually a physical wall!

In the 1980's, companies started building physical walls between their computers to stop hackers. Later, experts figured out how to build digital "firewalls" to protect data online!

Antivirus programs have huge "databases" that contain information about

thousands of viruses.

These databases are updated every day, so the antivirus can recognize new threats and protect your computer from the latest dangers.

On average, hackers attempt to attack someone's computer or online account

every 39 seconds.

With so many attacks happening every minute, it's clear that cybersecurity is more important than ever to protect your personal information!

Ethical hackers

use their hacking skills for good. They find flaws in systems and help fix them before bad hackers can exploit them.

Even satellites in space are targets for cyberattacks! In 2021, a group of hackers tried to

target satellites

that provide internet and communication services.

Did you know that your personal data—such as your name, birthday, and even the names of your pets—could be worth a lot of money to cybercriminals? They can

steal this info

and use it to commit fraud or identity theft.

The infamous 2017 WannaCry ransomware attack wreaked havoc on computers worldwide even though a software solution for it had been developed months before. WannaCry was able to spread because so many systems hadn't been updated, leaving them open to attack. That's a powerful reminder to

keep your system up to date!

WORDS TO KNOW

antivirus program a computer program that detects and disables computer malware.

artificial intelligence (AI) a type of computer program that can adapt and learn from data.

cyberattack gaining unauthorized digital access to someone else's computer.

cybercriminal someone who commits crimes using a computer.

cybersecurity a set of tools to keep computers safe from cyberattacks.

encryption using math to hide data in a way that can only be decoded with a key.

ethical hacker a computer expert who studies hackers and works to prevent malicious hacking.

firewall a cybersecurity program that is designed to keep malicious software from accessing a computer.

hacker cybercriminal who breaks into computer systems.

malware software that is designed to disrupt or damage a computer or computer system.

multifactor authentication (MFA) requiring two security-checking methods to access a site or device.

phishing sending emails with malicious or deceptive links.

software programs that run on a phone or computer.

spoofing sending emails pretending to be a different person or organization.

Trojan horse malicious software that inserts itself into programs on a computer by pretending to be something helpful.

INDEX

www.ingramcontent.com/pod-product-compliance
Lightning Source LLC
LaVergne TN
LVHW070859210926
848482LV00016B/975

* 9 7 8 0 7 1 6 6 6 7 2 6 1 *